LAB LOG BOOK

Log Book No.____

Continued From Log Book No. _______________________________________

Continued to Log Book No.____

Assigned To:

Name____

Signature__________________________________Date___________________

Date Issued_______________________By_______________________________

Phone_____________________________Email____________________________

Company / University / College ____________________________________

Department____

Address____

____

City__________________________State___________________Zip__________

Date Log Book Completed____

Number of Pages Filled In____

Notes:

// This Page Intentionally Blank //

Table of Contents	Page

Table of Contents	Page

Table of Contents	Page

Table of Contents	Page

Table of Contents	Page

Table of Contents	Page
Table of Contents	Page

Table of Contents

Table of Contents	Page

Table of Contents	Page

Project No. _______________

Book No. _______________   TITLE _______________

From Page No. _______________

To Page No. _______________

Witnessed & Understood by me,	Date	Invented by:	Date
		Recorded by:	

Project No. _______________

Book No. _______________   TITLE _______________

From Page No. _______

From Page No. _______   To Page No. _______

Witnessed & Understood by me,	Date	Invented by:	Date
		Recorded by:	

Project No. _______________

Book No. _______________   TITLE _______________

From Page No. _______________

To Page No. _______________

Witnessed & Understood by me,	Date	Invented by:	Date
		Recorded by:	

Project No. ______________________

Book No. ______________________   TITLE ______________________________________

From Page No. ______

From Page No. ______

To Page No. ________

Witnessed & Understood by me,	Date	Invented by:	Date
		Recorded by:	

Project No. ______________

Book No. ______________ TITLE ______________

From Page No. ______________

To Page No. ______________

Witnessed & Understood by me,	Date	Invented by:	Date
		Recorded by:	

Project No. _______________

Book No. _______________   TITLE _______________

From Page No. _______________

From Page No.

To Page No. _______________

| Witnessed & Understood by me, | Date | Invented by: | Date |

Recorded by:

Project No. _______________

Book No. _______________ TITLE _______________

From Page No. _______________

To Page No. _______________

Witnessed & Understood by me,	Date	Invented by:	Date
		Recorded by:	

Project No. ___________________

Book No. ___________________   TITLE ___________________________________

From Page No. ___________

To Page No. ___________

Witnessed & Understood by me,	Date	Invented by:	Date
		Recorded by:	

Project No. _______________

Book No. _______________   TITLE _______________

From Page No. _______

To Page No. _______

Witnessed & Understood by me,	Date	Invented by:	Date
		Recorded by:	

Project No. ______________

Book No. ______________   TITLE ______________

From Page No. ______________

To Page No. ______________

Witnessed & Understood by me,	Date	Invented by:	Date
		Recorded by:	

Project No. _______________

Book No. _______________   TITLE _______________

From Page No. _______________

To Page No. _______________

Witnessed & Understood by me,	Date	Invented by:	Date
		Recorded by:	

Project No. _______________

Book No. _______________   TITLE _______________

From Page No. _______________

To Page No. _______________

Witnessed & Understood by me,	Date	Invented by:	Date
		Recorded by:	

Project No. _______________

Book No. _______________   TITLE _______________

From Page No. _______________

To Page No. _______________

Witnessed & Understood by me, | Date | Invented by: | Date

Recorded by:

Project No. _____________________

Book No. _______________   TITLE _____________________________

From Page No. _________

From Page No. _________   To Page No. _________

Witnessed & Understood by me,	Date	Invented by:	Date
		Recorded by:	

Project No. _______________

Book No. _______________   TITLE _______________

From Page No. _______________

To Page No. _______________

Witnessed & Understood by me,	Date	Invented by:	Date
		Recorded by:	

Project No. ___________________

Book No. ___________________   TITLE ___________________

From Page No. ___________

To Page No. ___________

Witnessed & Understood by me,	Date	Invented by:	Date
		Recorded by:	

Project No. _______________

Book No. _______________   TITLE _______________

From Page No. _______________

To Page No. _______________

Witnessed & Understood by me,	Date	Invented by:	Date
		Recorded by:	

Project No. _______________

Book No. _______________   TITLE _______________

From Page No. _______________

To Page No. _______________

Witnessed & Understood by me,	Date	Invented by:	Date
		Recorded by:	

Project No. _______________

Book No. _______________   TITLE _______________

From Page No. _______________

To Page No. _______________

Witnessed & Understood by me,	Date	Invented by:	Date
		Recorded by:	

Project No. ____________________

Book No. ____________________   TITLE ____________________________________

From Page No. __________

To Page No. __________

Witnessed & Understood by me,	Date	Invented by:	Date
		Recorded by:	

Project No. ______________

Book No. ______________   TITLE ______________

From Page No. ______________

To Page No. ______________

Witnessed & Understood by me,	Date	Invented by:	Date
		Recorded by:	

Project No. ___________________

Book No. ___________________ TITLE ___________________________________

From Page No. ___________________

From Page No.

To Page No. ___________

Witnessed & Understood by me,	Date	Invented by:	Date
		Recorded by:	

Project No. _______________

Book No. _______________   TITLE _______________

From Page No. _______________

To Page No. _______________

Witnessed & Understood by me,	Date	Invented by:	Date
		Recorded by:	

Project No. _______________________

Book No. _______________________   TITLE _______________________

From Page No. _______________

To Page No. _______

Witnessed & Understood by me,	Date	Invented by:	Date
		Recorded by:	

Project No. _______________

Book No. _______________   TITLE _______________

From Page No. _______________

To Page No. _______________

Witnessed & Understood by me,   Date   Invented by: Date

Recorded by:

Project No. _______________

Book No. _______________   TITLE _______________

From Page No. _______________

To Page No. _______________

Witnessed & Understood by me,	Date	Invented by:	Date
		Recorded by:	

Project No. _______________

Book No. _______________   TITLE _______________

From Page No. _______

To Page No. _______

Witnessed & Understood by me,	Date	Invented by:	Date
		Recorded by:	

Project No. _______________

Book No. _______________   TITLE _______________

From Page No. _______________

To Page No. _______________

Witnessed & Understood by me,	Date	Invented by:	Date
		Recorded by:	

Project No. _______________

Book No. _______________   TITLE _______________

From Page No. _______________

To Page No. _______________

Witnessed & Understood by me, | Date | Invented by: | Date

Recorded by:

Project No. _______________

Book No. _______________   TITLE _______________

From Page No. _______________

From Page No. _______________

To Page No. _______________

Witnessed & Understood by me,	Date	Invented by:	Date
		Recorded by:	

Project No. _______________

Book No. _______________   TITLE _______________________________

From Page No. _______________

To Page No. _______________

Witnessed & Understood by me,   Date   Invented by: Date

Recorded by:

Project No. _______________

Book No. _______________   TITLE _______________________________

From Page No. _______

From Page No.

To Page No. _______

Witnessed & Understood by me,	Date	Invented by:	Date
		Recorded by:	

Project No. _______________

Book No. _______________   TITLE _______________________

From Page No. _______

To Page No. _______

Witnessed & Understood by me,	Date	Invented by:	Date
		Recorded by:	

Project No. ______________

Book No. ______________   TITLE ______________________________

From Page No. ______

To Page No. ______

Witnessed & Understood by me,	Date	Invented by:	Date
		Recorded by:	

Project No. ______________

Book No. ______________   TITLE ______________

From Page No. ______________

To Page No. ______________

Witnessed & Understood by me,	Date	Invented by:	Date
		Recorded by:	

Project No. ___________________

Book No. ________________   TITLE ___________________________

From Page No. _________

To Page No. _________

Witnessed & Understood by me,	Date	Invented by:	Date
		Recorded by:	

Project No. ________________

Book No. ________________   TITLE ________________

From Page No. ________________

To Page No. ________________

Witnessed & Understood by me, | Date | Invented by: | Date

Recorded by:

Project No. ___________________

Book No. ___________________ TITLE ___________________

From Page No. _______

From Page No. _______ To Page No. _______

Witnessed & Understood by me,	Date	Invented by:	Date
		Recorded by:	

Project No. _______________

Book No. _______________   TITLE _______________

From Page No. _______________

To Page No. _______________

Witnessed & Understood by me,	Date	Invented by:	Date
		Recorded by:	

Project No. _______________

Book No. _______________   TITLE _______________________________________

From Page No. _______

From Page No. _______

To Page No. _______

Witnessed & Understood by me,	Date	Invented by:	Date
		Recorded by:	

Project No. _______________

Book No. _______________   TITLE _______________

From Page No. _______________

To Page No. _______________

Witnessed & Understood by me,   Date   Invented by: Date

Recorded by:

Project No. _______________

Book No. _______________   TITLE _______________

From Page No. _______________

To Page No. _______________

Witnessed & Understood by me,	Date	Invented by:	Date
		Recorded by:	

Project No. _______________

Book No. _______________   TITLE _______________

From Page No. _______________

To Page No. _______________

Witnessed & Understood by me, | Date | Invented by: | Date

Recorded by:

Project No. ___________________

Book No. ___________________   TITLE ___________________________

From Page No. ___________

Project No.

Book No.

From Page No. ___________   To Page No. _________

Witnessed & Understood by me,	Date	Invented by:	Date
		Recorded by:	

Project No. _______________

Book No. _______________   TITLE _______________________________

From Page No. _______

To Page No. _______

Witnessed & Understood by me, | Date | Invented by: | Date

Recorded by:

Project No. ____________________

Book No. ____________________   TITLE ____________________

From Page No. __________

From Page No. __________

To Page No. __________

Witnessed & Understood by me,	Date	Invented by:	Date
		Recorded by:	

Project No. _______________

Book No. _______________   TITLE _______________

From Page No. _______

To Page No. _______

Project No. _______________

Book No. _______________   TITLE _____________________________

From Page No. _______

To Page No. _______

Witnessed & Understood by me,	Date	Invented by:	Date
		Recorded by:	

Project No. ___________________

Book No. ___________________   TITLE ___________________

From Page No. ___________________

To Page No. ___________________

Witnessed & Understood by me,	Date	Invented by:	Date
		Recorded by:	

Project No. _______________

Book No. _______________   TITLE _______________________

From Page No. _______

To Page No. _______

Witnessed & Understood by me,	Date	Invented by:	Date
		Recorded by:	

Project No. ______________

Book No. ______________   TITLE ______________

From Page No. ______________

To Page No. ______________

Witnessed & Understood by me,	Date	Invented by:	Date
		Recorded by:	

Project No. _______________

Book No. _______________   TITLE _______________

From Page No. _______

To Page No. _______

Witnessed & Understood by me,	Date	Invented by:	Date
		Recorded by:	

Project No. __________

Book No. __________   TITLE __________

From Page No. __________

To Page No. __________

Witnessed & Understood by me,   Date   Invented by: Date

Recorded by:

Project No. ___________________

Book No. ___________________   TITLE ___________________________________

From Page No. ___________

To Page No. ___________

Witnessed & Understood by me,   Date   Invented by: Date

Recorded by:

Project No. ______________

Book No. ______________   TITLE ______________

From Page No. ______________

From Page No. ______________

To Page No. ______________

Witnessed & Understood by me,	Date	Invented by:	Date
		Recorded by:	

Project No. ______________

Book No. ______________   TITLE ______________

From Page No. ______________

To Page No. ______________

| Witnessed & Understood by me, | Date | Invented by: | Date |

Recorded by:

Project No. ___________________

Book No. ___________________   TITLE ___________________________________

From Page No. ___________

From Page No. ___________

To Page No. ___________

Witnessed & Understood by me,	Date	Invented by:	Date
		Recorded by:	

Project No. _______________

Book No. _______________   TITLE _______________

From Page No. _______________

To Page No. _______________

Witnessed & Understood by me,	Date	Invented by:	Date
		Recorded by:	

Project No. _______________

Book No. _______________ TITLE _______________

From Page No. _______

To Page No. _______

Witnessed & Understood by me,	Date	Invented by:	Date
		Recorded by:	

Project No. _______________

Book No. _______________   TITLE _______________

From Page No. _______

To Page No. _______

Witnessed & Understood by me,	Date	Invented by:	Date
		Recorded by:	

Project No. ______________________

Book No. ______________________   TITLE ______________________

From Page No. ______________

From Page No. ______________   To Page No. __________

Witnessed & Understood by me,	Date	Invented by:	Date
		Recorded by:	

Project No. ___________________

Book No. ___________________   TITLE ___________________

From Page No. ___________

To Page No. ___________

Witnessed & Understood by me, | Date

Invented by: | Date

Recorded by:

Project No. ________________

Book No. ________________   TITLE ________________________________

From Page No. ________

To Page No. ________

Witnessed & Understood by me,	Date	Invented by:	Date
		Recorded by:	

Project No. ______________

Book No. ______________   TITLE ______________

From Page No. ______________

To Page No. ______________

Witnessed & Understood by me,   Date   Invented by: Date

Recorded by:

Project No. _______________

Book No. _______________   TITLE _______________

From Page No. _______

To Page No. _______

Witnessed & Understood by me,	Date	Invented by:	Date
		Recorded by:	

Project No. ___________________

Book No. ___________________   TITLE ___________________

From Page No. ___________

To Page No. ___________

Witnessed & Understood by me, | Date | Invented by: | Date

Recorded by:

Project No. _______________

Book No. _______________   TITLE _______________

From Page No. _______________

From Page No. _______________   To Page No. _______

Witnessed & Understood by me,	Date	Invented by:	Date
		Recorded by:	

Project No. _______________

Book No. _______________   TITLE _______________

From Page No. _______

To Page No. _______

Witnessed & Understood by me,	Date	Invented by:	Date
		Recorded by:	

Project No. _________________

Book No. _________________   TITLE _____________________________

From Page No. _________

To Page No. _________

Witnessed & Understood by me,	Date	Invented by:	Date
		Recorded by:	

Project No. _______________

Book No. _______________   TITLE _______________

From Page No. _______________

Witnessed & Understood by me,   Date   Invented by: Date

Recorded by:

To Page No. _______________

Project No. _______________

Book No. _______________   TITLE _______________________________

From Page No. _______

To Page No. _______

Witnessed & Understood by me,	Date	Invented by:	Date
		Recorded by:	

Project No. _______________

Book No. _______________   TITLE _______________

From Page No. _______________

To Page No. _______________

Witnessed & Understood by me, | Date | Invented by: | Date

Recorded by:

Project No. ______________

Book No. ______________   TITLE ______________________

From Page No. __________

To Page No. __________

Witnessed & Understood by me,	Date	Invented by:	Date
		Recorded by:	

Project No. _______________

Book No. _______________   TITLE _______________

From Page No. _______________

To Page No. _______________

Witnessed & Understood by me,   Date   Invented by: Date

Recorded by:

Project No. ___________________

Book No. ___________________   TITLE ___________________________________

From Page No. ___________

To Page No. ___________

Witnessed & Understood by me,	Date	Invented by:	Date
		Recorded by:	

Project No. _______________

Book No. _______________   TITLE _______________

From Page No. _______________

To Page No. _______________

| Witnessed & Understood by me, | Date | Invented by: | Date |

Recorded by:

Project No. ______________

Book No. ______________   TITLE ______________________________

From Page No. ______________

To Page No. ______________

Witnessed & Understood by me,	Date	Invented by:	Date
		Recorded by:	

CPSIA information can be obtained
at www.ICGtesting.com
Printed in the USA
LVHW031811250919
632254LV00015B/1769/P

9 781537 454795